Statistical Models
in Engineering

Statistical Models
in Engineering

Gerald J. Hahn & Samuel S. Shapiro
Research and Development Center, General Electric Company

JOHN WILEY & SONS
New York • Chichester • Brisbane • Toronto

ISBN 0 471 33915 6

Library of Congress Catalog Card Number: 67-12562
Printed in the United States of America
10 9

To Bea and Yevette for their help, patience, and understanding
and
To Adrienne, Susan, and Judith and Bonnie and Michael

SYSTEMS ENGINEERING AND ANALYSIS SERIES

In a society which is producing more people, more materials, more things, and more information than ever before, systems engineering is indispensable in meeting the challenge of complexity. This series of books is an attempt to bring together in a complementary as well as unified fashion the many specialties of the subject, such as modeling and simulation, computing, control, probability and statistics, optimization, reliability, and economics, and to emphasize the interrelationship between them.

The aim is to make the series as comprehensive as possible without dwelling on the myriad details of each specialty and at the same time to provide a broad basic framework on which to build these details. The design of these books will be fundamental in nature to meet the needs of students and engineers and to ensure they remain of lasting interest and importance.

Preface

This book is the outgrowth of many years of consulting with electrical, mechanical, chemical, industrial, and reliability engineers in widely divergent product areas in the General Electric Company and other organizations. In this work we have frequently felt the need for a detailed treatment, *directed at engineers*, on the use of statistical models to represent physical phenomena, and in this book we attempt to satisfy this need.

Our main purpose is to show the engineer responsible for a product whose performance is subject to chance fluctuations how to choose a reasonable statistical model and how to use this model in the evaluation of practical problems. Our choice of subjects is in line with this objective.

After some introductory comments in Chapter 1, the basic concepts of probability and distributional models are developed in Chapter 2. Chapters 3 and 4, which deal with continuous and discrete distributions, respectively, provide a detailed development of many models and the type of situations to which each is applicable. The relevance of the widely (and sometimes incorrectly) used normal distribution is considered, as is the applicability of the equally popular (and equally misused) exponential distribution in reliability problems. Alternate models, such as the gamma distribution, the beta distribution, the Weibull distribution and others are introduced and their use discussed. We have attempted to summarize the information in these chapters in a summary figure and table, which we hope will provide a fast reference guide.

The choice of a distribution among those presented in Chapters 3 and 4 is based principally on theoretical considerations. In some situations the underlying physical phenomena are insufficiently known for this to be possible, and a distributional fit must be evolved from the given data. Some versatile distributions for this purpose, known as Johnson and Pearson distributions, are discussed in Chapter 6.

Often the engineer is concerned with a system, ranging from a simple circuit to a complicated space vehicle, in which each component is subject

ix

to statistical variability. From knowledge of the system structure and the nature of the component behavior we draw conclusions about system performance. Because of its importance to engineering problems, we cover this problem in much greater detail than has been done to date in the literature accessible (and understandable) to the practicing engineer. In particular, an exact method, known as the transformation of variables, is described in Chapter 5. The approximate methods of generation of system moments (sometimes referred to as "statistical error propagation") and Monte Carlo simulation are considered in Chapter 7. The three methods are compared and the practical applicability of each is discussed. In the consideration of generation of system moments, expressions for evaluating the average, the spread, the symmetry, and the peakedness of the distribution of system performance are given and used to obtain distribution fits with the empirical models developed in Chapter 6. In the discussion of Monte Carlo simulation we include a description of methods for generating random values from the various models introduced in Chapters 3, 4, and 6 and also consider the number of simulations required to obtain a prespecified degree of precision.

The last chapter is devoted to methods of evaluating the adequacy of a chosen model and deals with both graphical and analytical procedures. The use of probability plots for estimating distribution parameters, as well as for evaluating the applicability of the model, is described for many of the distributions previously introduced (and not just for the normal distribution). Normal probability plots of data from various known distributions are shown to indicate the sensitivity of the method. In the second part of the chapter we present some new and powerful statistical tests for evaluating the assumption of normal and exponential distribution models, developed recently by one of us (S. S. Shapiro in conjunction with M. B. Wilk) and published previously only in statistical journals.

Except for indicating procedures for estimating parameters for many of the models and determining confidence intervals for some of these parameters, we include little of the material generally associated with the term "statistical inference." These subjects are excellently covered in many available books (see bibliography). Their omission here has hopefully allowed us to cover the subject of statistical models in some depth within the confines of a reasonably small volume.

We assume that the reader has the minimum undergraduate engineering mathematics and professional sophistication; however, no previous formal training in statistics is required. Those who have had an introductory course in probability and statistics will be acquainted with much of Chapter 2 and parts of Chapters 3 and 4 but should find most of the remainder of the book new. We attempt to give the theoretical justification necessary for the correct application of the methods, but in most cases derivation of results is either omitted or relegated to the appendices. A large number of illustrative examples are given, most of which have been adapted from actual problems; however, the name of the product or the actual numerical values have been changed "to protect the innocent."

Although the book is directed principally at practicing engineers, it has been used successfully as a supplementary text in an introductory graduate course in statistics for engineers at Union College, Schenectady, New York, with whose adjunct faculty we are both associated.

We are indebted to the Literary Executor of the late Sir Ronald A. Fisher, F.R.S., Cambridge, to Dr. Frank Yates, F.R.S., Rothamsted and to Messrs. Oliver & Boyd Ltd., Edinburgh, for permission to reprint Table III from their book *Statistical Tables for Biological, Agricultural, and Medical Research*. We are also indebted to Professor E. S. Pearson and the Biometrika Trustees for permission to reprint various tables from *Biometrika Tables for Statisticians*, Volume 1, by E. S. Pearson and H. O. Hartley; to Dr. Harold Chestnut and John Wiley and Sons for permission to reprint various figures from *Systems Engineering Tools*; to Dr. A. Hald and John Wiley and Sons for permission to reprint material from *Statistical Theory with Engineering Applications* and from *Statistical Tables and Formulas*; to Professors H. F. Dodge and H. G. Romig and John Wiley and Sons for permission to reprint material from *Sampling Inspection Tables*; to the Computational Laboratory, Harvard University, and the Harvard University Press for permission to reprint material from *Tables of the Cumulative Binomial Probability Distribution*; to the RAND Corporation and The Free Press, Macmillan, for permission to reprint material from *A Million Random Digits with 100,000 Normal Deviates*; and to Professors N. L. Johnson and E. S. Pearson for permission to reprint additional published and unpublished material.

It is a pleasure to acknowledge the substantial help we have had from many friends and associates in the preparation of this book. In particular, we are indebted to Messrs. Joseph J. Buckley, Charles B. Chandler, Ralph R. Duersch, Joel J. Fleck, Albert Fox, Paul Gunther, George E. Henry, John L. Jaech, William J. MacFarland, James E. Mulligan, James L. Myracle, Wayne B. Nelson, John Oldenkamp, Glenn M. Roe, Henry J. Scudder, III, and Hermann von Schelling for their constructive comments on preliminary versions of the manuscript; to Messrs. Harold Chestnut, Gilbert C. Dodson and Richard L. Shuey for their encouragement and guidance; to Professor Alfred L. Thimm and Dean William L. Weifenbach for arranging reproduction of the manuscript at Union College; to Mrs. Clara Doring for her excellent typing; and last (but not least) to Mrs. Gerald J. Hahn for her able preparation of the illustrations.

Schenectady, New York   GERALD J. HAHN
January 1967   SAMUEL S. SHAPIRO

Contents

Statistical Models
in Engineering

Chapter 1

Introduction

An important development in modern science and engineering is the study of systems in a probabilistic rather than a deterministic framework. The modern engineer, like his counterpart in many other fields, is becoming increasingly aware that deterministic models are inadequate for designing or evaluating the complex equipments of the twentieth century. Performance of supposedly identical systems differs because of many factors, such as component differences and fluctuations in the operating environment. Consequently, the engineer must be concerned with statistical models that describe these variations.

Once an understanding of such basic concepts is gained, it will seem natural to speak of the *statistical distribution* for output voltage of some system, rather than to specify only its design value or to talk about the *probability* that a component will not fail during a given time, instead of merely saying that it is not expected to fail. It will also seem reasonable to calculate circuit tolerances by statistical methods, rather than by the generally overconservative method of combining worst cases.

The application of statistical models in engineering is recent. The role of statistics in industry was limited prior to World War II. However, the widespread use of statistics in the war effort led to the rapid growth of the field in the immediate postwar period in such areas as quality control and communication theory. The present-day need for complex systems with high reliability has provided an even greater impetus to the more general use of statistics in industry.

As in other rapidly developing fields, education has lagged behind need, and only in very recent years have statistics courses found their way into the typical engineering curriculum. It is fair to say that even today probability and statistics in general, and the use of statistical models in particular, carry with them an aura of mystery for many practicing engineers. This should not be so, because statistics is no more difficult to comprehend than many other engineering disciplines.

1

1-1. Applications of Probability and Statistics

Probability and statistics are related fields. In problems in probability we make statements about the chances that various events will take place, based on an assumed model, whereas in problems in statistics we have some observed data and wish to determine a model that can be used to describe the data. Both situations arise frequently in engineering. For example, we may wish to predict the performance of a system of known design, before building, by assuming various models for the components that make up the system. This requires application of the techniques of probability theory. On other occasions test data on the performance of a system are given. Statistical techniques are then used to construct an appropriate model and to estimate its parameters. Once a model is obtained, it may, of course, be used to predict future performance.

Probability and statistics have many important applications in science and industry. Examples of problems—some of which are beyond the scope of the present book—in which these methods are used include the following:

1. Establishing sampling plans to control product quality economically.

2. Determining optimum redundancy of components in a complex system, such as a space vehicle, subject to weight restrictions.

3. Using past data to differentiate between potentially successful and unsuccessful computer programmers on the basis of an aptitude test.

4. Developing a sampling scheme that will ensure that a high proportion of bills sent to customers are error free.

5. Using readings taken at various tracking stations to determine a region in space within which we can be reasonably confident that some object is located.

6. Determining initial component burn-in and acceleration conditions that will eliminate a high proportion of failures due to manufacturing defects without harming good components.

7. Relating process variables to product performance to establish specification limits for control of future production.

8. Planning a multivariable development program to determine the effect of processing conditions on the properties of a chemical material, and interpreting the resulting data.

9. Estimating the fluctuations in performance of a large system, using knowledge of the variability and interrelationships of its elements.

10. Establishing an optimum strategy in competitive bidding when the alternate outcomes are subject to chance fluctuations, and, in general, making decisions in the face of uncertainties.

11. Finding the required level of power generation to ensure a high probability that the total demand for power will be met.

12. Comparing the consequences of alternate product warranty periods.

13. Establishing optimum product maintenance and replacement schedules.

14. Evaluating the effect of varying environments on the performance of a product.

15. Calibrating an instrument to obtain unbiased readings.

1-2. Scope of This Book

Our main concern is the use and manipulation of statistical models to represent engineering phenomena. Some important concepts of probability theory and the basic ideas of random variables and probability distributions are introduced in Chapter 2. Specific statistical models involving continuous and discrete random variables are described in Chapters 3 and 4. Methods for transforming random variables to represent specified physical situations are discussed in Chapter 5. Empirical frequency distributions useful in data fitting are presented in Chapter 6. Chapter 7 deals principally with two important procedures for drawing conclusions about system performance from component data—the method of generating system moments, and Monte Carlo simulation. The final chapter is concerned with evaluating the adequacy of a selected statistical model.

The emphasis is on providing insight into the nature and use of the models. Although we attempt to provide some theoretical justifications aimed at giving an understanding of basic concepts to the mature engineer, no claims of mathematical rigor are made. Complex derivations are generally omitted or relegated to an appendix. On the other hand, ample space is devoted to examples of the ideas under consideration. The level of discussion is thus aimed to be in line with our principal purpose—to provide working tools for practicing engineers who require the use of statistical models.

Material generally associated with statistical inference and statistical methods is treated lightly or omitted completely. We do give expressions for estimating the parameters for most of the important models and, in some cases, indicate the procedure for obtaining confidence intervals for measuring the precision of such estimates. However, little justification is given for these formulas. Also, there is no discussion of testing statistical hypotheses concerning distribution parameters or of more advanced techniques, such as the analysis of variance and least-squares regression. This material is well covered in many standard texts on statistics. To make locating them easier, we have included an extensive bibliography at the end of the book. Texts in such related areas as probability theory, design of engineering experiments, statistical quality control, and reliability methods are also mentioned. The availability of such material permits us to concentrate fully on our area of prime concern: statistical models in engineering.

Chapter 2

Probability and Random Variables

This chapter presents some basic concepts that are useful per se and that will be needed in the remainder of the book. The term *probability* will be interpreted in a number of ways. Since the theorems of probability can be best appreciated in terms of set theory, a brief review of this subject is given, followed by a statement of some basic laws of probability and examples of their application. The concept of a random variable and its associated probability distribution is presented next. We then consider various ways of summarizing the information about distributions and develop some rules about expected values and variances. The chapter also includes a discussion of distributions that involve more than one random variable. Because our objective is to present engineering applications, rather than mathematical theory, the treatment will be descriptive rather than theoretically rigorous. The reader is referred to some of the standard texts on probability and mathematical statistics, such as References 2-1 to 2-5 or those listed in the bibliography, for further details on the underlying theory.

2-1. Interpretations of Probability

In books on mathematical statistics the concept of probability is sometimes developed on a purely axiomatic basis. For our purposes it is more instructive to have first some intuitive insight into its meaning. Consequently, three conceptual interpretations are discussed below.

THE CLASSICAL (OR EQUALLY LIKELY) INTERPRETATION
OF PROBABILITY

If an event can occur in N equally likely and different ways, and if n of these ways have an attribute A, then the probability of the occurrence of A, denoted Pr *(A), is defined as n/N.*

Thus the probability of rolling a two with a perfect die is equal to $\frac{1}{6}$, for there are six equally likely outcomes, the numbers 1 to 6, of which only one has the value 2.

This definition is frequently inadequate for representing engineering situations. For example, it is not clear how it may be used to determine the probability of picking a defective unit from a process that in the past has given 73 defectives out of 10,000 units. In this case, making a defective unit and making a good one are not equally likely, and it is not clear what events are equally likely. Thus we require a second interpretation.

THE FREQUENCY (OR EMPIRICAL) INTERPRETATION OF PROBABILITY

If an experiment is conducted N times, and a particular attribute A occurs n times, then the limit of n/N as N becomes large is defined as the probability of the event A, denoted Pr (A).

Thus in the preceding example the probability of a defective is 73/10,000 or 0.0073 if we consider 10,000 to be a large number.

The frequency interpretation is the one that is used most generally by modern statisticians. However there are those who assert that even this definition is not general enough, because it does not cover the case in which little or no experimental evidence exists, and one's estimate concerning the outcome of an event is principally intuitive. This leads to the third definition.

THE SUBJECTIVE (OR MAN-IN-THE-STREET) INTERPRETATION OF PROBABILITY

The probability Pr (A) *is a measure of the degree of belief one holds in a specified proposition A.*

Under this interpretation, probability may be directly related to the betting odds one would wager on the stated proposition. The statement that the probability is 0.75 (or, equivalently, that the odds are three to one) that a specified change in the design will improve the performance of a given product involves the subjective definition, since it represents our degree of belief concerning the effect of the design change, based on engineering judgment or experience. Such a statement would be meaningless under the first two interpretations of probability. The classical interpretation is inadequate, because there is no reason to believe that it is just as likely that the product change will lead to an improvement as that it will not. The frequency interpretation is not applicable, because we do not have historical data on what proportion of the time this particular design change has succeeded in improving the performance of the given product. The subjective interpretation of probability is thus a broader concept than the classical and frequency interpretations, and it includes these other two interpretations.

In each of the preceding cases the probability of occurrence of an event is a number between zero, which corresponds to *no* chance that the event will occur, and one, which implies that the event *must* take place. Thus $0 \le \Pr(A) \le 1$. The basic rules for combining probabilities of events and many other results remain the same, irrespective of which of the above concepts is employed. And, although there is much dispute among mathematical statisticians about the appropriateness of the subjective interpretation as opposed to the other two, all three definitions have been included so that the engineer need not dogmatically limit himself to a single viewpoint.

2-2. Set-Theory Concepts

A formal calculus of probabilities can be developed in terms of sets. Therefore we shall give a brief, informal review of set theory.

A *set* is a finite or infinite collection of distinct objects or *elements* with some common distinguishing characteristic or characteristics. Examples of sets are the set of all engineers, the set of real numbers between zero and one, and the set of all persons who earn more than $10,000 a year and have red hair. Associated with the concept of a set is that of a *subset*—a partition of the set by some further characteristic that differentiates the members of the subset from the rest of the set. For example, the set of all engineers can be broken down into subsets composed of electrical engineers, mechanical engineers, and so on.

The set that contains all the elements under consideration is known as the *identity set* or *reference set* and will be denoted by the letter *I*. The letter *Z* will indicate a set with no elements, known as the *zero set*, *null set*, or *empty set*.

There are three basic operations in the manipulation of sets—the union operation, the intersection operation, and the complement operation.

THE UNION (OR) OPERATION

The *union* of two sets *A* and *B* is the set C_1, which is made up of the *distinct* elements that are contained in *A or B* or both and is indicated by

$$C_1 = A + B \tag{2-1}$$

or, in some books, by

$$C_1 = A \cup B. \tag{2-1a}$$

Because we consider distinct elements only, an element that is in both *A* and *B* is included only *once* in set C_1.

THE INTERSECTION (AND) OPERATION

The *intersection* of two sets A and B is the set C_2, made up of the elements that are common to both A *and* B, and is denoted by

$$C_2 = AB \tag{2-2}$$

or, in some books, by

$$C_2 = A \cap B \tag{2-2a}$$

or

$$C_2 = A \times B. \tag{2-2b}$$

THE COMPLEMENT (NOT) OPERATION

The complement of a set A is the set C_3, made up of all elements in the identity set *not* contained in A, and is denoted by

$$C_3 = \bar{A}. \tag{2-3}$$

It is convenient to illustrate set operations pictorially by means of a *Venn diagram* as shown in Figs. 2-1 and 2-2. The rectangles shown in the diagrams represents the identity set I. The area shown by the vertical lines represents a set A, and that shown by the horizontal lines represents a second set B. The total shaded region in Figs. 2-1 and 2-2a represents the union C_1 of the two sets—that is,

$$C_1 = A + B.$$

The unshaded region in Figs. 2-1 and 2-2a represents the complement $\bar{C}_1$ of the set C_1. Thus

$$\bar{C}_1 + C_1 = I.$$

The crosshatched region in Figs. 2-1 and 2-2b is the intersection of the two sets—that is,

$$C_2 = AB.$$

The sets

$$C_3 = \bar{A} \quad \text{and} \quad C_4 = \bar{B}$$

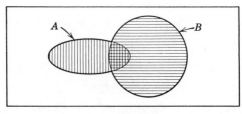

Fig. 2-1   Venn diagram showing two overlapping sets.

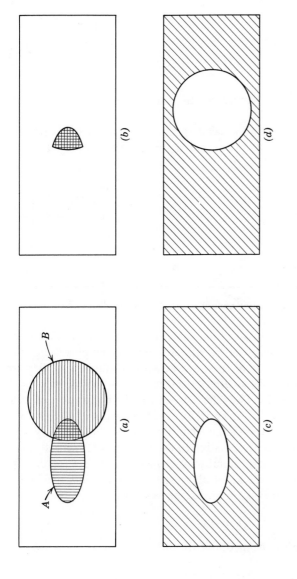

Fig. 2-2   Venn diagram elaborating on regions of Fig. 2-1: (a) $C_1 = A + B$ (shaded region), $\bar{C}_1$ (unshaded region); (b) $C_2 = AB$ (shaded region); (c) $C_3 = \bar{A}$ (shaded region); (d) $C_4 = \bar{B}$ (shaded region).

9

are represented in the Venn diagram by the shaded regions in Figs. 2-2c and 2-2d, respectively.

The following useful identities are stated without proof for an arbitrary set A:

$$A + Z = A; \qquad \bar{Z} = I; \qquad A + \bar{A} = I;$$
$$A + I = I; \qquad \bar{I} = Z; \qquad A + A = A;$$
$$AZ = Z; \qquad A\bar{A} = Z; \qquad AA = A; \qquad (2\text{-}4)$$
$$AI = A;$$

where, as before, Z and I represent the null set and the identity set, respectively.

The *number* of elements in the set A is denoted by $m(A)$ and is referred to as the *size* of the set A. Thus, for a set consisting of N elements, $m(I) = N$. Similarly, $m(Z) = 0$.

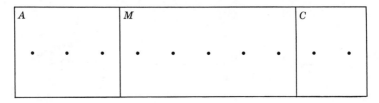

Fig. 2-3   Venn diagram representing nonprofessional employees at E. Z. Set Company.

The concept of the size of a set and some resulting rules are illustrated by the following example. The E. Z. Set Company employs ten non-professional workers. Three of these are assemblers (the set A), five are machinists (the set M), and two are clerks (the set C)—that is, $m(A) = 3$, $m(M) = 5$, and $m(C) = 2$. The reference set for the population of non-professional employees at the E. Z. Set Company consists of 10 elements; thus $m(I) = 10$. The corresponding Venn diagram is shown in Fig. 2-3. In this diagram the area corresponding to each of the three sets is shown as proportional to the relative number of elements in the set (or the size of the set). The set Q, made up of all workers who are *both* machinists and assemblers, does not contain any elements—that is,

$$Q = AM = Z,$$

where Z is the previously defined null set and

$$m(Q) = m(Z) = 0.$$

Two or more sets that contain no elements in common are said to be *mutually exclusive*. All the sets in this example are mutually exclusive.

The set F, consisting of all factory workers (assemblers and machinists), is the union of the sets A and M—that is,

$$F = A + M.$$

This set contains eight distinct elements, three from A and five from M. Thus

$$m(F) = m(A + M) = 3 + 5 = 8.$$

We note that in this case

$$m(A + M) = m(A) + m(M).$$

This result may be generalized to yield the following rule for determining the size of the union of two mutually exclusive sets A and B:

$$m(A + B) = m(A) + m(B) \qquad \text{if } AB = Z. \tag{2-5}$$

The size of the set of nonprofessionals other than assemblers—that is, $\bar{A}$—can be determined by examining the Venn diagram. Clearly,

$$\bar{A} = M + C.$$

Thus

$$m(\bar{A}) = m(M + C) = m(M) + m(C) = 5 + 2 = 7.$$

Equivalently,

$$m(\bar{A}) = m(I) - m(A) = 10 - 3 = 7.$$

Generalizing, we have a second rule for an arbitrary set A:

$$m(\bar{A}) = m(I) - m(A). \tag{2-6}$$

The E. Z. Set Company also employs eight full-time engineers (the set E), three full-time supervisors (the set S), and two individuals who are both engineers and supervisors (the set ES). The reference set of all employees of the E. Z. Set Company, shown by the Venn diagram in Figure 2-4, contains 23 elements.

In determining the size of the set of all professional employees (engineers and supervisors), it is not possible to use (2-5) because $ES \neq Z$. From the problem statement and examination of the shaded portion of Fig. 2-4 we see that $m(E + S) = 13$, rather than 15 as we would incorrectly have concluded by using (2-5). It is seen that in applying (2-5) to this situation, we count the elements that are common to both sets—that is, the elements in the set ES—twice. Thus we must subtract out the duplicate elements. This yields the following general rule for the size of the union of two sets A and B:

$$m(A + B) = m(A) + m(B) - m(AB). \tag{2-7}$$

Note that (2-5) is a special case of (2-7) in which $m(AB) = 0$.

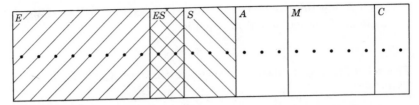

Fig. 2-4   Venn diagram for all employees of the E. Z. Set Company.

2-3. Probability and Set Theory

In probability theory the set elements are the outcomes of an experiment. Thus the reference set I is the set of all possible outcomes and is referred to as the *sample space* of the experiment. A particular outcome is known as a *sample point*. For example, in throwing a die, the sample space consists of the six sample points represented by the numbers 1, 2, 3, 4, 5, and 6. Thus $m(I) = 6$.

Frequently, we are concerned with a combination of possible outcomes, known as an *event*. An event is thus a subset of the sample space. Obtaining an even number in throwing the die is an example of an event—designated by A—made up of the sample points represented by the numbers 2, 4, and 6. Thus $m(A) = 3$. Similarly, the event B of "obtaining an odd number" consists of the three sample points represented by the numbers 1, 3, and 5. These two events exhaust the sample space. The individual outcomes may also be regarded as events. Since events are special types of sets, the methods described previously for obtaining new sets from existing ones also apply for events.

When all sample points are equally likely, as in the preceding example, finding the probability of some event A may be regarded as equivalent to comparing the relative size of the subset represented by the event A to that of the reference set I. Thus

$$\Pr(A) = \frac{m(A)}{m(I)}, \qquad (2\text{-}8)$$

and the probability of obtaining an even number in tossing a die is, as we all know, $\frac{3}{6}$ or 0.50.

We note that

$$\Pr(I) = \frac{m(I)}{m(I)} = 1 \qquad (2\text{-}8a)$$

and

$$\Pr(Z) = \frac{m(Z)}{m(I)} = 0, \qquad (2\text{-}8b)$$

as previously required. It is now seen that the relative size of some region in Figs. 2-3 and 2-4 is proportional to the probability of the outcome associated with the event the region represents.

When all points in the sample space are not equally likely, the procedure is the same, except that the sample points are no longer given *equal* weights. The specific assignment of weights depends upon the interpretation of probability and the underlying physical situation. The weight associated with a particular sample point might be proportional to its relative frequency of occurrence over many trials or to our degree of belief concerning the chances of its outcome on the next trial. The assignment is subject to the condition that the weights must add to unity when summed

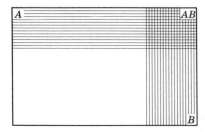

Fig. 2-5   Venn diagram for two independent events.

over all the points in the sample space. In our subsequent development of the laws of probability we shall generally assume, for the sake of simplicity, that we are dealing with equally likely sample points. However, the results hold irrespective of how the initial weights are assigned. Frequently it is more convenient to express probability as a percentage than as a proportion. Thus we shall sometimes say (for example) that the probability of the occurrence of some event is 50 per cent, rather than 0.50.

We now introduce the concept of independent events—a subject that will be elaborated on further in Section 2-5. Two events are said to be *independent* if the occurrence of one does not change the probability of the occurrence of the other. The outcomes of the consecutive tosses of two dice are examples of two independent events, as are the results of two random drawings from a bag containing five balls, each of which is different in color—assuming that the first ball drawn is replaced before the second drawing is made. If, instead, the first ball is *not* returned to the bag, the outcomes are no longer independent, because the selection on the first draw changes the contents of the bag for the second draw. Thus the probability that a particular ball will be picked on the second draw depends upon which ball has been selected on the first draw.

The Venn diagram of Fig. 2-5 shows two independent events, *A* and *B*. In this diagram the area representing the sets *A* and *B* has again been made

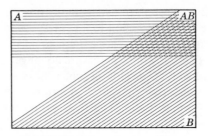

Fig. 2-6   Venn diagram for two nonindependent events.

proportional to the size of these sets. Note that the size of the set AB relative to that of the set A (the revised reference set *after* observing A) is the same as the size of the set B relative to the original reference set I. This is not the case when events are not independent; see Fig. 2-6. It also follows that if A and B are independent, then $\bar{A}$ and $\bar{B}$ are also independent.

2-4. *Probability Laws*

We shall now capitalize on the results of our study of set theory to obtain some important rules of probability.

Rule 1.   If $\Pr(A)$ and $\Pr(\bar{A})$ represent respectively the probabilities of the event A occurring and not occurring, then

$$\Pr(\bar{A}) = 1 - \Pr(A). \tag{2-6a}$$

This result follows directly from (2-6) and (2-8) and is suggested by the Venn diagram of Fig. 2-2c.

Rule 2.   If A and B are two *independent* events, then the probability that *both* A and B will happen, known as their *joint probability*, is the product of their respective individual probabilities—that is,

$$\Pr(A \text{ and } B) = \Pr(AB) = \Pr(A)\Pr(B) \qquad \text{if } A \text{ and } B \text{ are} \tag{2-9}$$
$$\text{independent.}$$

This result follows directly from our discussion in the preceding section. In fact, (2-9) is used frequently as the definition of two independent events A and B. In the above expression the joint probability, denoted by "and," is equivalent to the intersection operation of set theory.

Rule 3.   The probability of the joint occurrence of each of n *independent* events $A_1, A_2, \ldots, A_n$ is the product of their individual

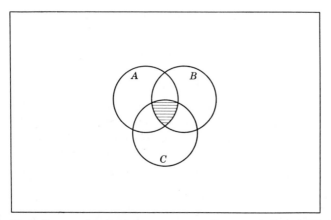

Fig. 2-7   Venn diagram showing joint occurrence of events A, B, and C by shaded region.

probabilities—that is,

$$\text{Pr}\,(A_1 \text{ and } A_2 \text{ and } \cdots A_n) = \text{Pr}\left(\prod_{i=1}^{n} A_i\right) = \text{Pr}\,(A_1)\,\text{Pr}\,(A_2)\cdots\text{Pr}\,(A_n).$$
$$(2\text{-}9a)$$

This result generalizes Rule 2 for the case of more than two events and is suggested for the joint occurrence of the three events A, B, and C by the shaded region of Fig. 2-7.

Rule 4.   If A and B are two *mutually exclusive events*—that is, if $m(AB) = 0$ and thus $\text{Pr}\,(AB) = 0$—then the probability that *one* of these two events will take place is given by the sum of their individual probabilities:

$$\text{Pr}\,(A \text{ or } B) = \text{Pr}\,(A + B) = \text{Pr}\,(A) + \text{Pr}\,(B). \qquad (2\text{-}5a)$$

(See total shaded region in Fig. 2-8.)

This result follows from (2-5) and (2-8). In the above expression the "or" operation is equivalent to the union operation of set theory.

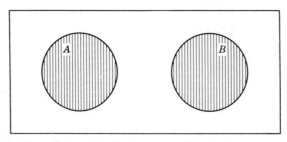

Fig. 2-8   Venn diagram showing two mutually exclusive events A and B.

Rule 5.   The probability of occurrence of one of n *mutually exclusive events* $A_1, A_2, \ldots, A_n$ is

$$\Pr(A_1 \text{ or } A_2 \text{ or } \cdots \text{ or } A_n) = \Pr\left(\sum_{i=1}^{n} A_i\right) = \sum_{i=1}^{n} \Pr(A_i). \qquad (2\text{-}5b)$$

This result is a direct generalization of Rule 4 for more than two events.

Rule 6.   If A and B are two events that are not necessarily mutually exclusive—that is, if $\Pr(AB) \neq 0$—then the probability that *at least one* of these two events will take place is given by the sum of their individual probabilities less their joint probability:*

$$\Pr(A \text{ and/or } B) = \Pr(A + B) = \Pr(A) + \Pr(B) - \Pr(AB). \qquad (2\text{-}7a)$$

(See shaded region in Fig. 2-2a.)

This result follows from (2-7) and (2-8). We note that, just as (2-5) is a special case of (2-7), so Rule 4 is a special case of Rule 6. Because the events are no longer mutually exclusive, the more general "and/or" operation replaces the "or" operation and becomes equivalent to the union operation of set theory.

We shall now extend Rule 6 to the case of more than two events. If we consider the event A and/or B as a single event, introduce a new event C and re-apply Rule 6, we obtain

$$\Pr(A \text{ and/or } B \text{ and/or } C)$$
$$= \Pr(A + B + C)$$
$$= \Pr(A) + \Pr(B) + \Pr(C) - \Pr(AB)$$
$$- \Pr(AC) - \Pr(BC) + \Pr(ABC). \qquad (2\text{-}10)$$

(See total shaded region of Fig. 2-9.)

A similar result is obtained for the probability that *at least one of n events* will take place. This leads to Rule 7.

Rule 7.   The probability that at least one of n events $A_1, A_2, \ldots, A_n$ will take place is

$$\Pr(A_1 \text{ and/or } A_2 \text{ and/or } \cdots \text{ and/or } A_n)$$
$$= \Pr\left(\sum_{i=1}^{n} A_i\right)$$
$$= \sum_{i=1}^{n} \Pr(A_i) - \sum_{j>i} \Pr(A_i A_j) + \sum_{k>j>i} \Pr(A_i A_j A_k)$$
$$- \sum_{l>k>j>i} \Pr(A_i A_j A_k A_l) + \cdots \qquad (2\text{-}11)$$

* The determination of joint probabilities when events are not independent will be discussed in the next section.

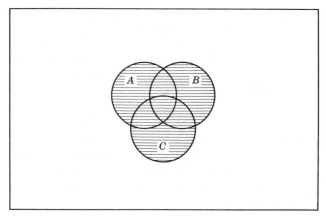

Fig. 2-9   Venn diagram showing occurrence of events A and/or B and/or C by shaded region.

Equivalently

$$\Pr(A_1 \text{ and/or } A_2 \text{ and/or } \cdots \text{ and/or } A_n) = 1 - \Pr(\bar{A}_1\bar{A}_2 \cdots \bar{A}_n); \quad (2\text{-}11a)$$

that is, the probability that at least one of the n events $A_1, A_2, \ldots, A_n$ will happen is one minus the joint probability that none of these events will take place. If the events are *independent*, it follows from (2-6a), (2-9a), and (2-11a) that

$$\Pr(A_1 \text{ and/or } A_2 \text{ and/or } \cdots \text{ and/or } A_n) = 1 - \prod_{i=1}^{n} [1 - \Pr(A_i)]. \quad (2\text{-}11b)$$

The following problem will illustrate some of the above rules.

It is desired to determine the reliability of a three-stage system, with parallel redundancy and probability of successful operation for each assembly as shown in Fig. 2-10. For example, the probability that assembly A will operate without failure is 0.9, the corresponding probability for assembly B is 0.8, and so on.

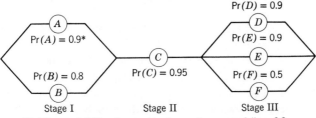

Fig. 2-10   System with parallel and series assemblies.   From H. Chestnut, *Systems Engineering Tools*, John Wiley and Sons, New York, 1965 (Fig. 6.1-1).

If the failure probabilities for each assembly and for each stage are independent of each other, what is the reliability—that is, the probability of successful operation—of the system?

Successful system operation requires that each of the three stages work successfully. Consequently, from Rule 3, the probability Pr (S) of successful system operation is

$$\text{Pr } (S) = \text{Pr } (\text{I}) \text{ Pr } (\text{II}) \text{ Pr } (\text{III}),$$

where Pr (I), Pr (II), and Pr (III) represent the probabilities of surviving stages I, II, and III, respectively. Now survival at stage I requires that either assembly A or assembly B or both work successfully. Therefore if Pr (A) designates the probability of successful operation for assembly A and so on, then, from Rule 6,

$$\begin{aligned}
\text{Pr } (\text{I}) &= \text{Pr } (A) + \text{Pr } (B) - \text{Pr } (AB) \\
&= (0.9) + (0.8) - (0.9)(0.8) = 0.98,
\end{aligned}$$

or equivalently, from (2-11b),

$$\begin{aligned}
\text{Pr } (\text{I}) &= 1 - [1 - \text{Pr } (A)][1 - \text{Pr } (B)] \\
&= 1 - [1 - 0.9][1 - 0.8] = 0.98.
\end{aligned}$$

Also

$$\text{Pr } (\text{II}) = \text{Pr } (C) = 0.95,$$

and, from (2-11b),

$$\begin{aligned}
\text{Pr } (\text{III}) &= 1 - [1 - \text{Pr } (D)][1 - \text{Pr } (E)][1 - \text{Pr } (F)] \\
&= 1 - [1 - 0.9][1 - 0.9][1 - 0.5] = 0.995.
\end{aligned}$$

Therefore

$$\text{Pr } (S) = (0.98)(0.95)(0.995) = 0.926,$$

which indicates that the chances are about 93 out of 100 that the system will operate successfully.

2-5. *Conditional Probability and Bayes' Theorem*

In many problems events are not independent. Consider, for example, the two-stage system shown in Fig. 2-11.

Assembly G is part of both stage I and stage II. Consequently, if this assembly fails during stage I, it is unavailable to operate successfully during stage II. Therefore, the survival probabilities for the two stages are not independent according to the definition of Section 2-3, because the probability of success in stage II depends upon what happens in stage I—specifically, whether or not assembly G survives stage I. In such situations we are interested in conditional probabilities. In particular, the *conditional probability* $\Pr(B \mid A)$ of an event B with respect to some other event A is the probability that B will occur, given that A has taken place. In the example $\Pr(\text{II} \mid \text{I})$ represents the probability of successful operation in stage II, given successful operation in stage I.

To understand the concept of conditional probability, consider the Venn diagram of Fig. 2-6. We are interested in the probability of the event B, given the occurrence of the event A. Once it is known that A has taken place, the set A replaces the set I as the sample space of interest. The size of the set AB, representing the joint occurrence of A and B relative to the new reference set, is given by $m(AB)/m(A)$. To change this expression

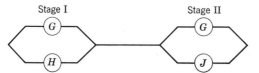

Stage I   Stage II

Fig. 2-11   Two-stage system with a nonindependent member. From H. Chestnut, *Systems Engineering Tools*, John Wiley and Sons, New York, 1965 (Fig. 6.1-2).

to a probability statement, we need only divide numerator and denominator by $m(I)$ according to (2-8), yielding

$$\Pr(B \mid A) = \frac{m(AB)/m(I)}{m(A)/m(I)} = \frac{\Pr(AB)}{\Pr(A)}. \tag{2-12}$$

We can now restate Rule 2 in a more general form, removing the requirement that the events be independent, as follows:

$$\Pr(A \text{ and } B) = \Pr(AB) = \Pr(A)\Pr(B \mid A). \tag{2-13}$$

In dealing with three events, A, B, and C, we similarly obtain

$$\Pr(A \text{ and } B \text{ and } C) = \Pr(ABC) = \Pr(A)\Pr(B \mid A)\Pr(C \mid AB), \tag{2-14}$$

where $\Pr(C \mid AB)$ denotes the conditional probability of C, given the occurrence of both A and B. Equation 2-14 can be generalized readily to obtain the joint probability of an arbitrary number of events.

These concepts are illustrated by the following example. Assume that 75 per cent of an inventory of transistors comes from vendor 1 and the remaining 25 per cent from vendor 2, and that 99 per cent of the units from vendor 1 and 90 per cent of those from vendor 2 give satisfactory performance. If we pick a transistor from inventory at random, what is the probability of selecting a unit that is made by vendor 1 and is also defective? What is the probability of selecting one that is defective, irrespective of vendor?

Let

$$A_1 = \text{event "transistor from vendor 1,"}$$
$$A_2 = \text{event "transistor from vendor 2,"}$$
$$B_1 = \text{event "good transistor,"}$$
$$B_2 = \text{event "defective transistor,"}$$
$$B_1 \mid A_1 = \text{event "transistor known to be from vendor 1 is good,"}$$

and so on. Then

$$\Pr(A_1) = 0.75, \qquad \Pr(A_2) = 0.25;$$
$$\Pr(B_1 \mid A_1) = 0.99, \qquad \Pr(B_2 \mid A_1) = 0.01;$$
$$\Pr(B_1 \mid A_2) = 0.90, \qquad \Pr(B_2 \mid A_2) = 0.10.$$

From (2-13) the probability that the transistor selected will be a defective unit from vendor 1 is

$$\Pr(A_1 B_2) = \Pr(A_1) \Pr(B_2 \mid A_1)$$
$$= (0.75)(0.01) = 0.0075.$$

Similarly, the probability that a defective unit from vendor 2 will be selected is

$$\Pr(A_2 B_2) = \Pr(A_2) \Pr(B_2 \mid A_2)$$
$$= (0.25)(0.10) = 0.025.$$

Because selection of a defective unit made by vendor 1 and selection of one made by vendor 2 are mutually exclusive events, it follows, from Rule 4, that the probability of selecting a defective unit, *irrespective of vendor*, is the sum of the probabilities of selecting a defective unit from vendor 1 and from vendor 2—that is,

$$\Pr(B_2) = \Pr(A_1 B_2) + \Pr(A_2 B_2)$$
$$= \Pr(A_1) \Pr(B_2 \mid A_1) + \Pr(A_2) \Pr(B_2 \mid A_2). \qquad (2\text{-}14a)$$

Therefore

$$\Pr(B_2) = 0.0075 + 0.0250 = 0.0325.$$

The generalized form of (2-14a) is frequently useful. Suppose the probability of the event B depends on some previous event that can occur in one of the n different ways $A_1, A_2, \ldots, A_n$. The unconditional probability $\Pr(B)$ can then be expressed as the sum of the conditional probabilities, weighted according to the probabilities of the respective A_i, where $i = 1, 2, \ldots, n$—that is,

$$\Pr(B) = \sum_{i=1}^{n} \Pr(B \mid A_i) \Pr(A_i), \qquad (2\text{-}15)$$

where the A_i are mutually exclusive and $\sum_{i=1}^{n} \Pr(A_i) = 1$.

An important law known as *Bayes' Theorem* follows directly from (2-15) and the concept of conditional probability. Bayes' Theorem provides a mechanism for combining the initial or *prior* probability concerning the occurrence of some event with related experimental data to obtain a revised or *posterior* probability. Consider the following situation:

1. There are n mutually exclusive events or states, $A_1, A_2, \ldots, A_n$. In the transistor example there were two such states: A_1, representing "transistor is from vendor 1," and A_2, representing "transistor is from vendor 2." Initial information is available that allows us to assign prior probabilities Pr (A_1), Pr (A_2), ... , Pr (A_n) to each of the A_i, subject to $\sum_{i=1}^{n}$ Pr $(A_i) = 1$. In the example Pr $(A_1) = 0.75$ and Pr $(A_2) = 0.25$.

2. The probability that some event B will take place depends on A_i in a known manner—that is, the probability of the occurrence of B, given A_1, is known; the probability of occurrence of B, given A_2, is known; and, in general, the conditional probabilities Pr $(B \mid A_i)$ for $i = 1, 2, \ldots, n$ are known. In the example, B might represent the selection of a defective transistor. Then Pr $(B \mid A_1) = 0.01$ and Pr $(B \mid A_2) = 0.10$.

3. We would like to know how the information that B has actually occurred alters the probabilities concerning each A_i—that is, we would like to evaluate the "after-the-event" probabilities Pr $(A_i \mid B)$ for $i = 1, 2, \ldots, n$. In the transistor example we might want to obtain a revised estimate of the probability that the selected transistor had been made by vendor 1 *after* learning that it was defective.

In summary, given (a) Pr (A_i) for $i = 1, 2, \ldots, n$ and (b) Pr $(B \mid A_i)$ for $i = 1, 2, \ldots, n$, we would like to find Pr $(A_i \mid B)$ for $i = 1, 2, \ldots, n$. From (2-13), the joint probability of A_i and B is

$$\Pr (A_i B) = \Pr (B) \Pr (A_i \mid B) = \Pr (A_i) \Pr (B \mid A_i),$$

from which we obtain

$$\Pr (A_i \mid B) = \frac{\Pr (A_i B)}{\Pr (B)} = \frac{\Pr (A_i) \Pr (B \mid A_i)}{\Pr (B)}.$$

Now, from (2-15),

$$\Pr (B) = \sum_{i=1}^{n} \Pr (B \mid A_i) \Pr (A_i),$$

which, when substituted in the preceding equation, yields the expression known as Bayes' Theorem, namely,

$$\Pr (A_i \mid B) = \Pr (A_i) \frac{\Pr (B \mid A_i)}{\sum_{i=1}^{n} \Pr (B \mid A_i) \Pr (A_i)} \qquad \text{for } i = 1, 2, \ldots, n. \quad (2\text{-}16)$$

The right-hand side of Bayes' equation is seen to consist of two terms: Pr (A_i), the prior probability, and Pr $(B \mid A_i)/\sum_{i=1}^{n}$ Pr $(B \mid A_i)$ Pr (A_i), the factor by which the prior probability is revised on the basis of the experimental data.

Returning to the transistor problem, the revised probability that the selected unit was made by vendor 1 *after* finding it to be defective is seen from (2-16) to be

$$\Pr(A_1 \mid B_2) = \Pr(A_1) \frac{\Pr(B_2 \mid A_1)}{\Pr(B_2 \mid A_1)\Pr(A_1) + \Pr(B_2 \mid A_2)\Pr(A_2)}$$

$$= (0.75) \frac{(0.01)}{(0.01)(0.75) + (0.10)(0.25)}$$

$$= 0.23.$$

Thus the prior probability of 0.75 that vendor 1 was the supplier is changed

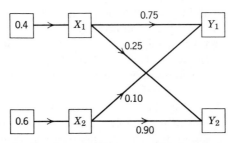

Fig. 2-12 Communication channel with two possible inputs, two possible outputs, and known transmission probabilities. From H. Chestnut, *Systems Engineering Tools*, John Wiley and Sons, New York, 1965 (Fig. 6.1-3).

to a posterior probability of 0.23, upon learning that the chosen unit was defective.

A further example is as follows: the transmission capacity of a communications channel depends on the probabilities of error within the channel. Consider a channel with two possible inputs, X_1 and X_2, and two possible outputs, Y_1 and Y_2. Assume that 40 per cent of the time the channel initially receives an X_1 and 60 per cent of the time it receives an X_2. There is a 0.75 probability of correctly transmitting an X_1 as Y_1. The probability that an input X_1 will incorrectly yield Y_2 as an output is then 0.25. Similarly, for a signal initially received as X_2, the probabilities of transmitting Y_2 and Y_1 are 0.90 and 0.10, respectively. The situation is represented diagrammatically in Fig. 2-12.

In a given situation an output of Y_1 is obtained. What is the probability that X_1 was the original input?

The prior probabilities are

$$\Pr(X_1) = 0.4, \qquad \Pr(X_2) = 0.6;$$

also

$$\Pr(Y_1 \mid X_1) = 0.75 \qquad \text{and} \qquad \Pr(Y_1 \mid X_2) = 0.10.$$

Then, from (2-16),

$$Pr\,(X_1 \mid Y_1) = Pr\,(X_1)\,\frac{Pr\,(Y_1 \mid X_1)}{Pr\,(Y_1 \mid X_1)\,Pr\,(X_1) + Pr\,(Y_1 \mid X_2)\,Pr\,(X_2)}$$

$$= 0.4\,\frac{(0.75)}{(0.75)(0.4) + (0.10)(0.6)} = 0.833.$$

Thus the information that Y_1 was the output changes the probability that X_1 was the input from 0.4 to 0.833.

In the preceding examples, the prior probabilities can presumably be obtained from knowledge of the physical situation or analysis of past data using one of the first two interpretations of probability at the beginning of this chapter. In many problems it is not possible to obtain such objective prior probabilities, and in such cases many statisticians believe that Bayes' Theorem is not applicable. In using the subjective interpretation of probability, however, we are not limited in this regard, since the prior probabilities can then be based upon personal judgment. Bayes' Theorem is in fact such a crucial tool in the analysis of situations involving subjective probabilities that *all* procedures involving such probabilities are frequently referred to as Bayesian methods. For further details see References 2-6 and 2-7.

2-6. Random Variables and Probability Functions

THE CONCEPT OF A RANDOM VARIABLE

We have seen in the preceding discussion that in using probability theory we need to determine the size of sets and subsets representing events of interest in relation to the size of the reference set or *sample space*. Up to now this has been done by collecting the sample points associated with a particular event and adding together their associated probabilities to obtain the probability of the event under consideration. In considering more complex situations, we need introduce a new concept—that of a random variable.

A *random variable* or *variate* is a function defined on a sample space. For example, say we are tossing two dice. The sample space then consists of the 36 possible pairs of outcomes: (1, 1), (1, 2), . . . , (6, 6), where the two members represent the results of the first and second tosses respectively. The *sum* of the outcomes for each pair of tosses is a random variable, because it is a function defined for every point in the sample space. The exact relationship between the points in the sample space and the values of the random variable is shown in Table 2-1. For example, the random variable takes on the value 2 for the sample point (1, 1), the value 3 for

Table 2-1   Relationship Between Sample Points and Random Variable Defined by Sum of Values in Tossing Two Dice

Sample Point	Corresponding Value of Random Variable	Sample Point	Corresponding Value of Random Variable
1, 1	2	4, 1	5
1, 2	3	4, 2	6
1, 3	4	4, 3	7
1, 4	5	4, 4	8
1, 5	6	4, 5	9
1, 6	7	4, 6	10
2, 1	3	5, 1	6
2, 2	4	5, 2	7
2, 3	5	5, 3	8
2, 4	6	5, 4	9
2, 5	7	5, 5	10
2, 6	8	5, 6	11
3, 1	4	6, 1	7
3, 2	5	6, 2	8
3, 3	6	6, 3	9
3, 4	7	6, 4	10
3, 5	8	6, 5	11
3, 6	9	6, 6	12

the sample point (1, 2), and so on. Thus, although the sample space consists of 36 points, the random variable takes on only the 11 integral values from 2 to 12. This is because a number of the points in the sample space map into the same value of the random variable. More formally, let y_1 and y_2 represent the results of the first and second tosses respectively. Then the random variable of interest is the *function*

$$x = y_1 + y_2.$$

Many other random variables could be defined on the same sample space. Some examples are the following:

1. The average value of the two tosses—that is, the random variable

$$x_1 = \frac{y_1 + y_2}{2},$$

which takes on the values 1, $\frac{3}{2}$, 2, $\frac{5}{2}$, 3, $\frac{7}{2}$, 4, $\frac{9}{2}$, 5, $\frac{11}{2}$, 6.

2. The square of the summed value of the two tosses—that is, the random variable

$$x_2 = (y_1 + y_2)^2,$$

which takes on the values 4, 9, 16, 25, 36, 49, 64, 81, 100, 121, 144.

3. The value of the first toss only—that is, the random variable

$$x_3 = y_1,$$

which takes on the values 1, 2, 3, 4, 5, 6. This random variable is equivalent to the points in the sample space in the single-die tossing example of Section 2-3.

The concept of a random variable provides a mechanism for mapping the qualitative results of an experiment on to a quantitative scale. Consider, for example, a manufacturing process that makes five devices daily. Each device is classified either as good (G) or as defective (D). The sample space then consists of all possible sequenced combinations of good and defective units—that is, the points (G, G, G, G, G), (G, G, G, G, D), (G, G, G, D, G), ..., (D, D, D, D, D), where the first value in each group indicates whether the first device is good (G) or defective (D), the second value refers to the status of the second device, and so on. It can be seen that the sample space consists of the total of $2^5 = 32$ points. A random variable of interest on this sample space is the number of defective units made during a particular day. This random variable takes on the values 0, 1, 2, 3, 4, and 5 and involves the following mapping:

Point in Sample Space	Corresponding Value of Random Variable
(G, G, G, G, G)	0
(G, G, G, G, D)	1
(G, G, G, D, G)	1
.	.
.	.
.	.
(D, D, D, D, D)	5

The introduction of the random variable in the preceding example thus transforms the qualitative points in the sample space to an integer-valued variable of physical importance.

The two preceding examples are concerned with sample spaces containing only a finite number of values. Frequently when dealing with counted data, such as the number of flaws in a piece of material, one observes random variables that can, at least theoretically, take on *any* integral value. In this case, the sample space is said to be made up of a countable infinity of points. A sample space involving either a finite number or a countable infinity of elements is said to be *discrete*. A *discrete random variable* is one that can take on only a countable number of values.

Further examples of discrete random variables are the number of phone calls received at an exchange in a given hour, the number of equipments arriving at a repair facility in a given day, the number of baseball games won by the New York Mets in a season, and the number of failures observed on a system performance test.

A second type of random variable is a *continuous* one. In contrast to a discrete random variable, a continuous random variable may take on any value in one or more intervals. Continuous random variables result when we are dealing with measured, rather than counted, data. Examples of continuous random variables are the time to failure of an electron tube, the height of an individual from a particular group, the noise level in decibels of an appliance, and the proportion of defective material in a steel casting.

PROBABILITY FUNCTION AND CUMULATIVE DISTRIBUTION OF A
DISCRETE RANDOM VARIABLE

Since a random variable is a function defined on a sample space, we can associate probabilities with the values of the random variable. This is done by a probability function. Thus the *probability function* of a discrete random variable x is an expression $p(x_i)$ that gives the probability associated with all possible values of the random variable—that is, $p(x_i) = \Pr(x = x_i)$, for $i = 1, 2, \ldots$ Note that

$$p(x_i) \geq 0 \qquad \text{for all } x_i. \tag{2-17}$$

The probability function assigns a weight to each value of the random variable based upon the probability content of the subset of the sample space associated with that value. In the two-dice example the probability of each value of the random variable x representing the sum of the results of two tosses is obtained, according to Rule 5 of Section 2-3, by adding the probabilities of appropriate points in the sample space. Because each of the 36 points is equally likely and their total probability must add to 1, each point has associated with it a probability of $\frac{1}{36}$.

The probability function shown in Table 2-2 and sketched in Fig. 2-13 is thus obtained. For example, $p(4)$, the probability that $x = 4$, equals

*Table 2-2   Probability Function for Sum of Values
Obtained in Tossing Two Dice*

Value of Random Variable (x_i)	2	3	4	5	6	7	8	9	10	11	12
$p(x_i)$	$\frac{1}{36}$	$\frac{2}{36}$	$\frac{3}{36}$	$\frac{4}{36}$	$\frac{5}{36}$	$\frac{6}{36}$	$\frac{5}{36}$	$\frac{4}{36}$	$\frac{3}{36}$	$\frac{2}{36}$	$\frac{1}{36}$

$3(\frac{1}{36})$, for there are three points in the sample space—(1, 3), (2, 2) and (3, 1)—for which $y_1 + y_2 = 4$. Equivalently, the probability function may be written more compactly as follows:

$$p(x_i) = \begin{cases} \dfrac{x_i - 1}{36} & x_i = 2, 3, \ldots, 6, \\[2mm] \dfrac{13 - x_i}{36} & x_i = 7, 8, \ldots, 12. \end{cases} \qquad (2\text{-}18)$$

In many practical problems, instead of being interested in the probability that a random variable x takes on a *specific* value x_i, we want to know the

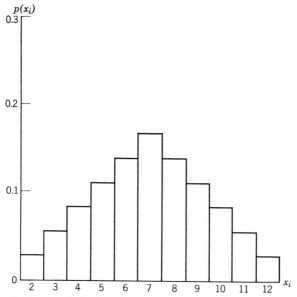

Fig. 2-13   Probability function for sum of values in tossing two dice.

probability that x is *less than or equal to* x_i. This probability is given by the cumulative distribution function. In particular, the function $F(x_i)$, which gives the probability of obtaining a value smaller than or equal to some value x_i of the discrete random variable x, is known as the *cumulative distribution function* or *cumulative distribution* or *distribution function* of that random variable. $F(x_i)$ can be obtained by summing the values of the probability function over those points in the sample space for which the random variable takes on a value less than or equal to x_i—that is,

$$\Pr(x \le x_i) = F(x_i) = \sum_{x \le x_i} p(x_i). \qquad (2\text{-}19)$$

Table 2-3   *Cumulative Distribution Function for Sum of Values Obtained in Tossing Two Dice*

Value of Random Variable (x_i)	< 2	2	3	4	5	6	7	8	9	10	11	≥ 12
$F(x_i)$	0	$\frac{1}{36}$	$\frac{3}{36}$	$\frac{6}{36}$	$\frac{10}{36}$	$\frac{15}{36}$	$\frac{21}{36}$	$\frac{26}{36}$	$\frac{30}{36}$	$\frac{33}{36}$	$\frac{35}{36}$	1

Clearly

$$0 \leq F(x_i) \leq 1 \qquad \text{for all } x_i$$

and

$$F(x_i) \geq F(x_j) \qquad \text{for } x_i \geq x_j.$$

In the two-dice tossing problem, the cumulative distribution function is as shown in Table 2-3 and sketched in Fig. 2-14.

This cumulative distribution could also be written in equation form, in a manner similar to (2-18).

The complement of the distribution function gives the probability that the random variable *exceeds* a specified value—that is,

$$\Pr (x > x_i) = 1 - F(x_i). \qquad (2\text{-}20)$$

Also

$$\sum p(x_i) = 1, \qquad (2\text{-}21)$$

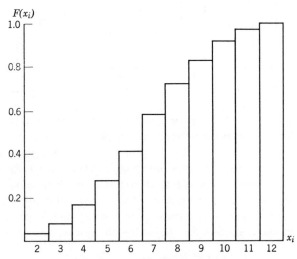

Fig. 2-14   Cumulative distribution function for sum of values in tossing two dice.

because the probability associated with the totality of points in the sample space is 1, and the definition of the random variable requires a mapping of every point in the sample space.

PROBABILITY DENSITY FUNCTION AND CUMULATIVE DISTRIBUTION FOR A CONTINUOUS RANDOM VARIABLE

As in the case of a discrete random variable, the *cumulative distribution function* (or *cumulative distribution* or *distribution function*) $F(x_i)$ of a continuous random variable x gives the probability that x takes on a value equal to or less than some specified x_i—that is,

$$F(x_i) = \Pr(x \leq x_i).$$

Consequently, the probability that x has a value between x_1 and x_2 is the difference between the values of the cumulative distribution evaluated at these two points—that is,

$$\Pr(x_1 < x \leq x_2) = F(x_2) - F(x_1). \tag{2-22}$$

Similarly,
$$\Pr(x > x_i) = 1 - F(x_i).$$

It follows that if $F(x_i)$ is the cumulative distribution of a continuous random variable x, then

$$\lim_{x_i \to -\infty} F(x_i) = F(-\infty) = 0,$$

$$\lim_{x_i \to \infty} F(x_i) = F(\infty) = 1,$$

$$F(x_i) \geq 0 \qquad \text{for all } x_i, \tag{2-22a}$$

and
$$F(x_i) \geq F(x_j) \qquad \text{if } x_i > x_j.$$

As an example, consider the time to decay, t, for a radioactive particle. Assume that from theoretical considerations we know that the probability that a particle will *survive* beyond time t_i is $e^{-\lambda t_i}$, where λ is a known constant—that is,

$$\Pr(t > t_i) = 1 - F(t_i) = e^{-\lambda t_i}.$$

Consequently, the cumulative distribution function for decay time is

$$\Pr(t \leq t_i) = F(t_i) = 1 - e^{-\lambda t_i}. \tag{2-22b}$$

The plot of this cumulative distribution function with $\lambda = 0.1$ is shown in Fig. 2-15. It may be easily verified that $F(t_i)$ meets the requirements specified by (2-22a). Note that the sample space for t includes all non-negative times t_i—that is, it is theoretically possible for the decay to occur at *any* instant of time.

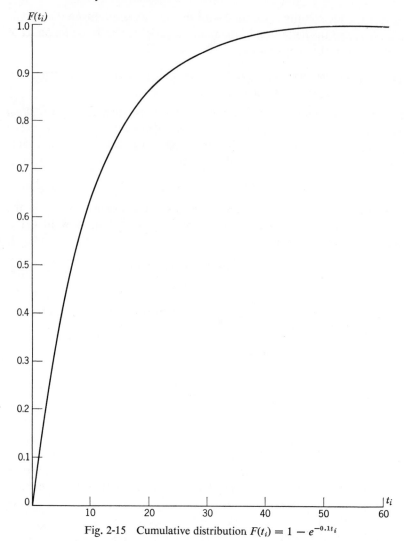

Fig. 2-15   Cumulative distribution $F(t_i) = 1 - e^{-0.1t_i}$

The probability that decay will occur during the time interval (t_1, t_2) may be obtained from (2-22) as

$$\Pr(t_1 < t \le t_2) = F(t_2) - F(t_1) = (1 - e^{-\lambda t_2}) - (1 - e^{-\lambda t_1})$$
$$= e^{-\lambda t_1} - e^{-\lambda t_2}. \tag{2-22c}$$

For example, if time is expressed in hours and $\lambda = 0.1$, then by (2-22b) the probability of decay during the first two hours is

$$\Pr(t \le 2) = F(2) = 1 - e^{-0.1(2)} = 0.181,$$

and from (2-22c) the probability of decay some time during the fifth hour is

$$\Pr(4 < t \le 5) = F(5) - F(4) = [1 - e^{-0.1(5)}] - [1 - e^{-0.1(4)}] = 0.063.$$

Because the sample space for a continuous random variable contains an uncountable infinite number of points, the probability associated with any particular value of a continuous random variable is zero. Thus, for example, the left-hand side of (2-22) could have been written as $\Pr(x_1 < x < x_2)$ or $\Pr(x_1 \le x \le x_2)$.

For a discrete random variable the probability function $p(x_i)$ was defined as the probability associated with the value x_i. Such a direct definition is clearly no longer meaningful for a continuous random variable. Instead, we use the definition of the cumulative distribution function to define the *probability density function* (or *probability density* or *density function* or *density*) $f(x)$ of a continuous variate x as follows:

$$f(x) = \lim_{\Delta x \to 0} \frac{\Pr(x_i \le x \le x_i + \Delta x)}{\Delta x} = \frac{d}{dx}[F(x)]. \qquad (2\text{-}23)$$

The term probability density function arises from the fact that although $f(x)$ is not a probability, $f(x)\,\Delta x$ approximately equals $\Pr(x_i \le x \le x_i + \Delta x)$ provided Δx is small enough.

The probability density function for time to decay in the previous problem is thus

$$f(t) = \frac{d}{dt}[1 - e^{-\lambda t}] = \lambda e^{-\lambda t}, \qquad t > 0. \qquad (2\text{-}24)$$

Equation 2-24 is known as the exponential probability density function and will be discussed further in Chapter 3 and in later sections of this chapter.

The probability that x will take on a value less than or equal to x_1 may be found from the probability density function as follows:

$$\Pr(x \le x_1) = F(x_1) = \int_{-\infty}^{x_1} f(y)\,dy. \qquad (2\text{-}25)$$

Similarly,

$$\Pr(x > x_2) = 1 - F(x_2) = \int_{x_2}^{\infty} f(y)\,dy \qquad (2\text{-}25a)$$

and

$$\Pr(x_1 < x \le x_2) = F(x_2) - F(x_1)$$

$$= \int_{-\infty}^{x_2} f(y)\,dy - \int_{-\infty}^{x_1} f(y)\,dy = \int_{x_1}^{x_2} f(y)\,dy. \qquad (2\text{-}26)$$

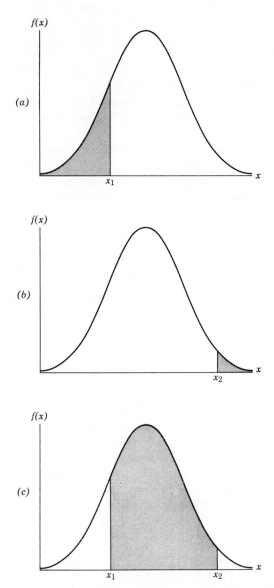

Fig. 2-16   Probability density function $f(x)$ showing relative area corresponding to (a) probability that x is less than x_1; (b) probability that x is more than x_2; and (c) probability that x is between x_1 and x_2.

Probability for a continuous random variable may thus be interpreted in terms of relative area under the curve defined by the probability density function. In particular, for a continuous random variable x with probability density function $f(x)$:

1. The probability that x will be less than x_1 is represented by the relative area under $f(x)$ to the left of x_1 (see Fig. 2-16a).

2. The probability that x will be more than x_2 is represented by the relative area under $f(x)$ to the right of x_2 (see Fig. 2-16b).

3. The probability that x will be between x_1 and x_2 is represented by the relative area under $f(x)$ between x_1 and x_2 (see Fig. 2-16c).

It also follows from (2-22a) and (2-23) that

$$\int_{-\infty}^{\infty} f(x)\, dx = F(\infty) - F(-\infty) = 1 \tag{2-27}$$

and

$$f(x) \geq 0 \qquad \text{for all } x. \tag{2-27a}$$

Thus, in choosing a probability density function for a continuous random variable, we must limit ourselves to those nonnegative-valued functions whose integral equals unity over the specified sample space or range of variation. The equivalent requirements for a discrete random variable are given by (2-21) and (2-17).

FURTHER COMMENTS AND DEFINITIONS

In the preceding discussion we have used the letter x (or t) to designate the random variable and have denoted the *value* that the random variable takes on by x_i (or t_i). Two other common conventions are (a) to designate the random variable by X and its value by x, and (b) to use heavy print to indicate the random variable and light print for its value. For the sake of simplicity, especially when dealing with continuous variates, we often use the letter x to designate *both* the random variable and its value. The exact meaning should be clear from the context. Also, the term *distribution* is used in a generic sense to denote probability function or probability density function. Thus, for example, we might say that time to decay follows an exponential distribution or that it is an exponentially distributed variate.

We have attempted to show that it is convenient to use the concept of a random variable to describe chance events. Associated with each random variable is a distribution, which, when known, describes the probabilistic behavior of the underlying system. One or more constants, known as *parameters*, which provide the proper location, scaling, and shape, are generally included in the definition of the distribution. The constant λ in the particle-decay problem is an example of a scale parameter. In many problems the *type* of distribution is known from theoretical or engineering considerations, but it will be necessary to determine the parameters of the distribution from the available data. We generally differentiate between a known parameter value and an estimate thereof, based on observed data, by placing a hat ($\hat{}$) above the estimate. Thus $\hat{\lambda}$ denotes an estimate of the parameter λ of the exponential distribution.

In dealing with random variables and statistical models, we frequently deal with the selection of a *random sample* to estimate the value of one or more parameters, to evaluate the adequacy of the model, and so on. A random sample is a selection of units from a population such that every unit in the population has a fixed and known probability of being picked. In this book we restrict ourselves to random samples for which every unit in the population has an *equal* probability of inclusion. Such random samples may be obtained from a fixed population of size N by assigning a different number from 1 to N to each unit and using a table of uniform random numbers (see Chapter 7) for selecting the units to be included in the sample.

A tabulation of the observed values of the random sample, generally arranged in grouped form (or frequency cells), especially for continuous variates, will be referred to as an *empirical frequency distribution* or *frequency distribution*. A plot of the frequency distribution, with rectangles proportional in height to the class frequencies in the sample on the ordinate against the corresponding values of the random variable on the abscissa, is called a *histogram*. The next two chapters are devoted to a discussion of a wide variety of distributions used to represent continuous and discrete random variables.

A function of a random variable is also a random variable. In the two-dice problem, for example, the square of the sum of the results of two tosses, the function $u(x) = x^2$, is a random variable. In Chapter 5 procedures for determining the probability density of *functions* of random variables are discussed.

2-7. Central Values of Distributions

Frequently an engineer wants to summarize the information about the distribution of a random variable by a few descriptive values. At other times the form of the underlying distribution is not known, but it is still of interest to calculate summary measures from available data.

One of the most important single descriptive values is the point around which the distribution is centered, known as a measure of *central tendency*. Three different ways of describing this central point will be indicated.

THE EXPECTED VALUE, OR MEAN

The best known measure of central tendency is the *expected value*, more frequently called the *arithmetic mean*, or sometimes just *the average** or *the mean*. It is defined as

$$E(x) = \begin{cases} \int_{-\infty}^{\infty} x f(x)\, dx, & \text{if } x \text{ is a continuous random variable with probability density function } f(x), \\[2ex] \sum_i x_i\, p(x_i), & \text{if } x \text{ is a discrete random variable with probability function } p(x_i). \end{cases} \quad (2\text{-}28)$$

We may think of the probability function or probability density function as a procedure for assigning relative weights to the values of the random variable. Then the expected value may be regarded as the center of gravity of the distribution, since it is *that point* around which the sum of the distance to the left times the probability weight exactly balances out the corresponding sum of weighted values to the right. This is illustrated in Figs. 2-17 and 2-18 for the following two examples. In tossing a single die the expected value, using the discrete form of (2-28), is

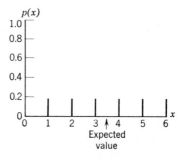

Fig. 2-17   Probability function for result in single toss of die, indicating expected value.

$$E(x) = \sum_{x_i=1}^{6} x_i\, p(x_i) = 1\left(\frac{1}{6}\right) + 2\left(\frac{1}{6}\right)$$

$$+ 3\left(\frac{1}{6}\right) + 4\left(\frac{1}{6}\right) + 5\left(\frac{1}{6}\right) + 6\left(\frac{1}{6}\right)$$

$$= 3.5.$$

(See Fig. 2-17.)

In this case, the random variable never takes on its expected value.

Similarly, the expected value of the sum of the outcomes in tossing two dice (see Table 2-2) is

$$E(x) = \sum_{x_i=2}^{12} x_i\, p(x_i) = 2\left(\frac{1}{36}\right) + 3\left(\frac{2}{36}\right) + 4\left(\frac{3}{36}\right) + 5\left(\frac{4}{36}\right) + 6\left(\frac{5}{36}\right)$$

$$+ 7\left(\frac{6}{36}\right) + 8\left(\frac{5}{36}\right) + 9\left(\frac{4}{36}\right) + 10\left(\frac{3}{36}\right) + 11\left(\frac{2}{36}\right) + 12\left(\frac{1}{36}\right)$$

$$= 7$$

(see Fig. 2-18).

* However, the term average is also used for other measures of central tendency.

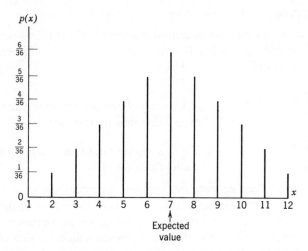

Fig. 2-18 Probability function for sum of results in tossing two dice, indicating expected value.

The expected value of the exponentially distributed random variable t, representing particle decay time, is

$$E(t) = \int_{-\infty}^{\infty} t f(t) \, dt = \int_{0}^{\infty} t\lambda e^{-\lambda t} \, dt = \frac{1}{\lambda}, \qquad (2\text{-}29)$$

using integration by parts. Thus the mean decay time equals the reciprocal of the distribution parameter λ.

The expected value is used often in statistical analysis and will be encountered frequently in the remainder of this book.

THE MEDIAN

Another measure of central tendency is the mid-point or *median* of the distribution. For a continuous probability density function $f(x)$, the median is the point z such that

$$\int_{-\infty}^{z} f(x) \, dx = 0.5. \qquad (2\text{-}30)$$

Thus the median is that value of the random variable that has exactly one half of the area under the probability density function to its left and one half to its right (see Fig. 2-19). The median for a discrete random variable is defined similarly, except that the integral is replaced by a summation over the values of the random variable.

In the particle-decay problem the median z is known as the half-life, because about one-half of a large number of particles would decay by this time. Its value is found by solving

$$\int_0^z \lambda e^{-\lambda t}\, dt = 0.5.$$

This yields $z = (\ln 2)/\lambda$. Thus the median of an exponentially distributed variate is a little over two thirds its expected value.

The median often is an appropriate measure of central tendency for random variables that are not symmetrically distributed. For example, the distribution of personal income for a group of families is perhaps "better represented" by the median than by the arithmetic mean. This is

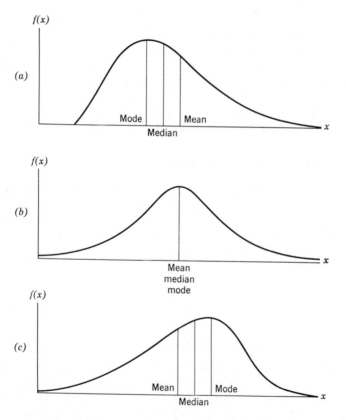

Fig. 2-19   Relationship between mean, median, and mode for three probability density functions. From H. Chestnut, *Systems Engineering Tools*, John Wiley and Sons, New York, 1965 (Fig. 6.3-1).

because the median is not as sensitive to a small number of extreme observations—such as families with very high incomes—as is the mean.

THE MODE

The *mode* is a third measure of central tendency. For a discrete random variable the mode is that value of the random variable that has the highest probability. The mode of a continuous variate is the value associated with the maximum of the probability density function (if a unique maximum exists). An exponentially distributed variate has its mode at the origin, since the maximum value of

$$f(t) = \lambda e^{-\lambda t}, \qquad t \geqslant 0, \qquad \lambda > 0$$

is $t = 0$.

COMPARISON OF MEASURES OF CENTRAL TENDENCY

We note from the preceding discussion that the arithmetic mean is sensitive to extreme observations, the median is less affected, and the mode is *not* influenced at all by such observations. The approximate relationship among these three values is shown graphically for three hypothetical probability densities in Fig. 2-19. It is seen that for a single-peaked symmetric distribution, the three measures coincide. This in general is not the case for asymmetric and multipeaked distributions.

THE DETERMINATION OF DATA MEAN, MEDIAN, AND MODE

The foregoing discussion has dealt with measures of central tendency when the mathematical form of the distribution is known. Frequently we are interested in estimating these values when all that is given is the values of n observations.* In such cases the following rules apply:

Calculation of the Data Mean.   The data mean, denoted by $\bar{x}$, is calculated as

$$\bar{x} = \frac{\sum\limits_{i=1}^{n} x_i}{n}, \tag{2-31}$$

where x_i, $i = 1, 2, \ldots, n$, are the values for the n data points.

Calculation of the Data Median.   Rank the observations according to their magnitude. Then (a) if n is odd, the median is the value of the $[(n + 1)/2]$th ranked observation; (b) if n is even, the median is the mean of the $(n/2)$th and the $[(n/2) + 1]$th ranked observations.

* When the data are grouped in a frequency distribution the midpoint of each frequency class is used represent all the observations in that class.

Calculation of the Data Mode.   Determine the value of the data point that occurs most frequently. In some situations it might be reasonable to group the data into frequency classes of equal length. In such a case the mode is taken as the center point of the frequency class that contains the largest number of observations.

The following example illustrates these procedures. Ten 5-μF capacitors were placed on test. The change in capacitance on each unit after 100 hours is shown in Table 2-4.

Table 2-4   Change in Capacitance After 100 Hours of Test for Ten 5-μF Capacitors

Unit No.	Change in Capacitance (μF)	Unit No.	Change in Capacitance (μF)
1	−0.10	6	−0.06
2	−0.01	7	0.00
3	0.00	8	−0.08
4	+0.02	9	−0.03
5	−0.15	10	+0.01

The mean capacitance change is obtained from the data by using (2-31):

$$\bar{x} = \frac{\sum_{i=1}^{10} x_i}{10} = -\frac{0.40}{10} = -0.04$$

To determine the median, the observations are ranked according to increasing magnitude, as shown in Table 2-4a. Because n equals 10, an

Table 2-4a   Ranked Values of Change in Capacitance After 100 Hours of Test for Ten 5-μF Capacitors

Rank	Unit No.	Change in Capacitance (μF)
1	5	−0.15
2	1	−0.10
3	8	−0.08
4	6	−0.06
5	9	−0.03
6	2	−0.01
7 and 8	3 and 7	0.00
9	10	+0.01
10	4	+0.02

even number, the median is the average of the fifth and sixth observations—that is, $(n/2) = 5$ and $[(n/2) + 1] = 6$. The value of the fifth
observation is $-0.03 \mu F$ and the value of the sixth observation is $-0.01 \mu F$.
Their mean is $-0.02 \mu F$, which is the median change in capacitance, as
calculated from the given data.

The mode of the given data—the most frequently occurring value of
capacitance change—is $0 \mu F$, although for such a small number of
observations the mode is not a very meaningful value.

Before continuing our discussion of procedures for summarizing information about distributions, we shall briefly consider the problem of
determining expected values in some more complicated situations.

2-8. Expected Value of a Function of a Random Variable

As indicated previously, if $h(x)$ is a function of the random variable x,
then $h(x)$ is also a random variable. The procedure for determining the
distribution of $h(x)$ is discussed in Chapter 5. Instead of having to know
the complete distribution of $h(x)$, it is often sufficient to find only its
expected value. This could be done by obtaining the distribution of the
new variable and taking its expected value. However, determining the
distribution of a function of a random variable can sometimes be very
tedious. In this section we shall therefore consider situations in which
we may find the expected value of $h(x)$ directly.

The following example illustrates the problem. Each time an equipment
breaks down the manufacturer is liable for a dollar penalty equal to the
square of the number of hours, t, required for repair. Say that t is an
exponentially distributed random variable and the parameter λ is the
repair rate. What is the expected value of the penalty per breakdown?

The function $h(t)$ relating repair time to dollar penalty is

$$h(t) = t^2,$$

where t has the probability density function

$$f(t) = \lambda e^{-\lambda t}, t \geq 0.$$

We desire to determine $E[h(t)]$.

The expected value of a mathematical function $h(x)$ is defined by

$$E[h(x)] = \begin{cases} \int_{-\infty}^{\infty} h(x) f(x)\, dx & \text{if } x \text{ is a continuous random variable} \\ & \text{with probability density function } f(x), \\ \\ \sum_i h(x_i)\, p(x_i) & \text{if } x \text{ is a discrete random variable with} \\ & \text{probability function } p(x_i). \end{cases}$$

(2-32)

Equation 2-32 is a generalization of (2-28), with the random variable x replaced by the new variable $h(x)$.

In the problem just stated the expected value of the penalty per breakdown is

$$E(t^2) = \int_0^\infty t^2 \lambda e^{-\lambda t}\, dt = \frac{2}{\lambda^2}. \qquad (2\text{-}33)$$

For example, if $\lambda = 0.1$ per hour, the expected dollar value of the penalty per breakdown is $2/(0.1)^2$ or 200.

The following important results also follow from (2-32):

1. The expected value of a constant c equals the constant itself—that is,

$$E(c) = c \qquad (2\text{-}34)$$

—because if x is a continuous random variable,

$$E(c) = \int_{-\infty}^\infty c\, f(x)\, dx = c \int_{-\infty}^\infty f(x)\, dx = c,$$

and if x is a discrete variate,

$$E(c) = \sum_{i=1}^n c\, p(x_i) = c \sum_{i=1}^n p(x_i) = c.$$

2. The expected value of a constant times a random variable equals the constant times the expected value of the random variable—that is,

$$E(cx) = c\, E(x) \qquad (2\text{-}35)$$

—because, from (2-32),

$$E(cx) = \int_{-\infty}^\infty cx\, f(x)\, dx = c \int_{-\infty}^\infty x\, f(x)\, dx = c\, E(x)$$

for the continuous case and

$$E(cx) = \sum_{i=1}^n cx_i\, p(x_i) = c \sum_{i=1}^n x_i\, p(x_i) = c\, E(x)$$

for the discrete case.

Sometimes, instead of dealing with a function of a single random variable x, we are concerned with a function of a number of random variables. For this situation the following two results, to be proved in Section 2-11, are applicable.

3. The expected values of the sum of n random variables $x_1, x_2, \ldots, x_n$ is the sum of their individual expected values—that is,

$$E\left[\sum_{j=1}^n x_j\right] = \sum_{j=1}^n E(x_j). \qquad (2\text{-}36)$$

4. More generally, for any linear combination of random variables

$$E\left[\sum_{j=1}^{n} b_j x_j\right] = \sum_{j=1}^{n} [b_j\, E(x_j)], \qquad (2\text{-}36a)$$

where $x_1, x_2, \ldots, x_n$ are random variables and $b_1, b_2, \ldots, b_n$ are constants.

Thus, in the equipment-breakdown example, if a manufacturer is liable for a delay penalty (x_1) plus half the material repair cost (x_2) and one quarter of the labor repair cost (x_3) each time the equipment breaks down, the expected total cost per breakdown is

$$E(x_1 + 0.5x_2 + 0.25x_3) = E(x_1) + 0.5E(x_2) + 0.25E(x_3).$$

If these three components have expected values of $1000, $500, and $300, respectively, then the expected value or total dollar cost per breakdown is

$$\$1000 + 0.5(\$500) + 0.25(\$300) = \$1325.$$

We note that (2-34), (2-35), and (2-36) are special cases of (2-36a). These results hold irrespective of the distribution of each of the random variables, the specific form of which need not even be known, and are valid whether or not the random variables are independent (see Section 2-11 for the definition of independent random variables).

The preceding discussion will be extended in Chapter 7 to a consideration of approximate methods for obtaining the expected value of more complicated functions of random variables. We shall now return to our review of ways of summarizing information about distributions.

2-9. Other Descriptive Measures of Distributions

MOMENTS OF A DISTRIBUTION

In addition to the central value, we frequently wish to describe distribution spread, symmetry, and peakedness. These characteristics may be summarized by the *moments* of the distribution. To simplify the notation μ_k' is used to designate $E(x^k)$. Thus μ_1' denotes the expected value $E(x)$ of the random variable x.* μ_k' is also known as the kth *raw moment* of the distribution or the kth *moment about zero*, or about the origin.

The kth *moment about the mean* or *central moment* is defined as

$$\mu_k = E[x - \mu_1']^k = \begin{cases} \displaystyle\int_{-\infty}^{\infty} (x - \mu_1')^k f(x)\, dx & \text{if x is continuous with probability density function $f(x)$,} \\[2ex] \displaystyle\sum_i (x_i - \mu_1')^k\, p(x_i) & \text{if x is discrete with probability function $p(x_i)$.} \end{cases}$$

$$(2\text{-}37)$$

* However, $E(x)$ is also frequently designated by μ.

In the terminology of the preceding section we may think of $(x - \mu_1')^k$ as the function $h(x)$ of the random variable x. Thus (2-37) is a special case of (2-32).

A distribution is completely specified once all its moments are known.* However, many distributions can be adequately described by the first four moments, and discussion will be limited to these moments. In addition to providing useful descriptive measures, the first four moments play an important role in fitting empirical distributions (see Chapter 6) and in approximating the distribution of a random variable (see Chapter 7).

The first central moment is always zero, that is

$$\mu_1 = 0,$$

for

$$\mu_1 = E[x - \mu_1'] = E(x) - E(\mu_1') = \mu_1' - \mu_1' = 0.$$

THE VARIANCE AND STANDARD DEVIATION

The second moment about the mean is a measure of dispersion. It is known as the *variance* and, according to (2-37), is defined as

$$\mu_2 = \text{Var}(x) = E[x - \mu_1']^2$$

$$= \begin{cases} \int_{-\infty}^{\infty} (x - \mu_1')^2 f(x)\, dx & \text{for the continuous and case and} \\ \sum_i (x_i - \mu_1')^2 p(x_i) & \text{for the discrete case.} \end{cases} \quad (2\text{-}38)$$

Equivalently,

$$\text{Var}(x) = \mu_2' - (\mu_1')^2. \quad (2\text{-}39)$$

The justification of (2-39) is as follows:

$$\text{Var}(x) = E(x - \mu_1')^2$$
$$= E[x^2 - 2x\mu_1' + (\mu_1')^2]$$
$$= E(x^2) - 2\mu_1' E(x) + (\mu_1')^2.$$

Therefore

$$\text{Var}(x) = \mu_2' - (\mu_1')^2.$$

The variance of an exponentially distributed random variable t is obtained from (2-29), (2-33), and (2-39) as

$$\text{Var}(t) = \frac{2}{\lambda^2} - \left(\frac{1}{\lambda}\right)^2 = \frac{1}{\lambda^2}. \quad (2\text{-}40)$$

* The use of the moment-generating function to obtain all desired moments of a distribution is described in the books on statistical theory listed in the bibliography.

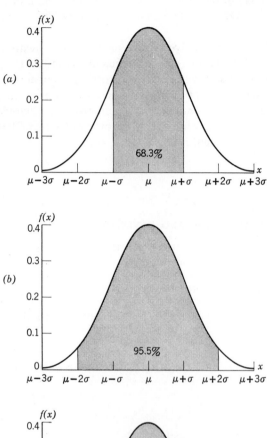

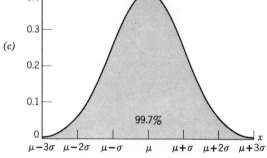

Fig. 2-20 Area under normal distribution in following intervals: (a) $\mu \pm \sigma$; (b) $\mu \pm 2\sigma$; (c) $\mu \pm 3\sigma$.

The square root of the variance is known as the *standard deviation* and is denoted by the symbol σ. Thus for an exponential variate $\sigma = 1/\lambda$. The standard deviation is expressed in the same units as the original variate. As seen in Chapter 3, for a random variable following a normal distribution 68.3 per cent of the probability is within $\pm 1\sigma$ around the mean and 95.5 and 99.7 per cent of the probability are within the $\mu \pm 2\sigma$ and $\mu \pm 3\sigma$ ranges, respectively (see Fig. 2-20). In dealing with other variates, it is sometimes useful to obtain bounds that may be very conservative by using a theorem known as *Tchebychev's Inequality*. This states that for *any distribution* with finite mean and variance, *at least* $[1 - (1/k^2)]$ times 100 per cent of the probability is in the range $\pm k\sigma$ around the mean. Thus for most distributions at least 75 per cent of the area under the distribution curve is within $\mu \pm 2\sigma$, and at least 88.9 per cent is within $\mu \pm 3\sigma$.

SKEWNESS

The third moment about the mean is related to the asymmetry or *skewness* of a distribution and, according to (2-37), is defined as

$$\mu_3 = E(x - \mu_1')^3. \tag{2-41}$$

A useful formula is

$$\mu_3 = E(x - \mu_1')^3 = \mu_3' - 3\mu_2'\mu_1' + 2(\mu_1')^3, \tag{2-42}$$

the derivation of which is analogous to that for (2-39).

A unimodal (i.e. a single peaked) distribution with $\mu_3 < 0$ is said to be skewed to the left—that is, it has a left "tail" (see Fig. 2-19c). If $\mu_3 > 0$, the distribution is skewed to the right (see Fig. 2-19a). For symmetric distributions $\mu_3 = 0$ (see Fig. 2-19b).

The quantity

$$\sqrt{\beta_1} = \frac{\mu_3}{(\mu_2)^{3/2}} \tag{2-43}$$

measures the skewness of the distribution relative to its degree of spread. This standardization allows us to compare the symmetry of two distributions whose scales of measurement differ; $\sqrt{\beta_1}$ is also sometimes denoted by α_3.

For an exponentially distributed variate it was previously found that

$$\mu_1' = \frac{1}{\lambda}$$

and

$$\mu_2' = \frac{2}{\lambda^2}.$$

From (2-32)

$$\mu_3' = E(t^3) = \int_0^\infty t^3 \lambda e^{-\lambda t}\, dt = \frac{6}{\lambda^3}.$$

Thus, using (2-42),

$$\mu_3 = \frac{6}{\lambda^3} - 3\left(\frac{2}{\lambda^2}\right)\left(\frac{1}{\lambda}\right) + 2\left(\frac{1}{\lambda}\right)^3 = \frac{2}{\lambda^3}. \tag{2-44}$$

The positive value of μ_3 (since λ is greater than zero) shows that the exponential distribution is skewed to the right. From (2-43)

$$\sqrt{\beta_1} = \frac{2/\lambda^3}{(1/\lambda^2)^{3/2}} = 2. \tag{2-45}$$

Thus the value of $\sqrt{\beta_1}$ for an exponentially distributed random variable is independent of the value of the distribution parameter λ.

KURTOSIS

The fourth moment about the mean is related to the peakedness—also called *kurtosis*—of the distribution and is defined as

$$\mu_4 = E(x - \mu_1')^4. \tag{2-46}$$

It can easily be shown that

$$\mu_4 = E(x - \mu_1')^4 = \mu_4' - 4\mu_3'\mu_1' + 6\mu_2'(\mu_1')^2 - 3(\mu_1')^4. \tag{2-47}$$

The quantity

$$\beta_2 = \frac{\mu_4}{\mu_2^2}, \tag{2-48}$$

also known as α_4, is a relative measure of kurtosis. In Fig. 2-21 two probability densities are shown: the uniform or rectangular distribution, and the bell-shaped curve that represents the normal distribution. The values of β_2 for these two distributions are 1.8 and 3.0, respectively.

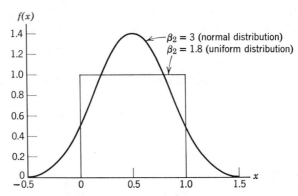

Fig. 2-21 Comparison of normal and uniform probability density functions.